Valerio Travi

Advanced Technologies
Building in the Computer Age

Birkhäuser – Publishers for Architecture
Basel • Boston • Berlin

Translation into English: Bertrand Colson, Bologna

A CIP catalogue record for this book is available from the Library of Congress, Washington D.C., USA.

Deutsche Bibliothek Cataloging-in-Publication Data

Travi, Valerio:
Advanced technologies : building in the computer age / Valerio Travi. [Transl. into Engl.: Bertrand Colson]. - Basel ; Boston ; Berlin : Birkhäuser, 2001
    (The IT revolution in architecture)
    ISBN 3-7643-6450-5

Original edition:
(Universale di Architettura 97, collana fondata da Bruno Zevi; La Rivoluzione Informatica, sezione a cura di Antonino Saggio).
© 2001 Testo & Immagine, Turin

© 2001 Birkhäuser – Publishers for Architecture, P.O. Box 133, CH-4010 Basel, Switzerland.
Member of the BertelsmannSpringer Publishing Group.
Printed on acid-free paper produced of chlorine-free pulp. TCF ∞
Printed in Italy
ISBN 3-7643-6450-5

9 8 7 6 5 4 3 2 1   http://www.birkhauser.ch

Contents

1. The Context

Over the last thirty years, with the gradual application of artificial intelligence inside buildings, the worlds of construction and technical systems have witnessed revolutionary changes. The first automatic systems in the 60s consisted of simple switchboards which could be programmed and controlled at a distance. Since then, spurred by the energy crisis in the 70s, we have progressed to complex automated systems networks linking various apparatuses. Developments in electronic technology and growing collaboration between the fields of information and telecommunication technologies have led us to make the most of the possibilities offered by centralized control systems. In some buildings, the different automated systems are now interconnected, from the system for energy management to those for vertical transport, security, or fire prevention.

Networks which convey digital information (computer data, telephone signals, and signals for environmental control and security) and usually function independently of each other, can now be integrated and thereby optimized.

The process of integrating systems consists not only in linking various cables. The aim is to enable and boost interaction between the various devices and apparatuses, from the PC to the automated heating system. Various electronic technologies now adopted the world over must be applied and combined according to new principles and concepts. The introduction of complex system engineering in building construction has led to the appearance of new terms and abbreviations such as Building Automation, Computer Integrated Building, and domotique. These vary from one country to the next according to what installations are most required. Some terms may refer to particular systems contained within the building per se. For example, Building Automation indicates that an automated system is in charge of all technical installations, or Home Systems, that security functions, domestic appliances and other house installations are all managed by various automated systems. Others, however, refer to services accessible from within the building by means of advanced systems. For instance, Shared Tenant

Services is a term used mainly in the US to indicate that a building is equipped with devices which allow the user access to various telecommunication and management services. These various terms do not necessarily indicate actual differences in terms of system engineering. Rather, they are connected with the appearance on the market of new product configurations or with the merger of producers looking to establish a new brand name.

Building automation is divided into two different areas in which essentially similar systems are installed and configured in entirely different ways so that different kinds of products are marketed. These two categories are building automation and home automation. Clearly, the main difference between these two kinds of automation is one of scale. Building automation concerns large constructions which accommodate great numbers of people and are generally used for business and corporate-sector activities. On the other hand, home automation concerns homes, detached or terraced houses or flats; spaces where, though a small number of people is involved, the system must meet the extremely diversified needs of each family member. This same idea holds true for the functions carried out by the automation systems. In the large office building, where the system's tasks and functions are standardized and recurrent, centralized management is possible and appropriate. Practically speaking, the automated system is required to transform the building into a machine and guarantee its best possible operation. In the home, however, the system's tasks are less clearly defined and what most matters is reliability and overall security.

There is also a significant difference in system-user interfaces. In the case of building automation, given the complexity of the installations and high number of control points and devices, professional operators (so-called *building operators*) are responsible for the system and must be familiar enough with its operation to be able to interpret available data. In the case of residential automation, though the house may be empty during the holidays or when the owner is at work, the system must be much more flexible and its interface much more "user-friendly," since all family members may want to use it, even older persons or young children.

The difference in scale also affects the choice of the technician responsible for installing the system. Whereas engineering firms and producers deal with large commercial buildings, it is difficult to imagine individuals turning to the same companies for their private homes. They will rather turn to a qualified electrician or installer. Above all, they seek the assistance of the many security operators already active in the private sector. However, new solutions made possible, for example, by the Internet point to more fluid infrastructures and open up the field to professional figures from other sectors.

1.1 In-depth Analysis

FACTORS DETERMINING THE DEVELOPMENT
OF ADVANCED TECHNOLOGY BUILDING CONSTRUCTION

1960s *organizational efficiency*
1970s *functional efficiency*
1980s *work-environment quality*
1990s *adaptation capability*
2000   *widening connections and multimedia links*

Progress in applied information and telecommunication technologies has played a fundamental role in the development of the intelligent building. Technological progress has led to a constant expansion of the services and functions a building may provide and perform. The first automated systems appeared in the 1980s and consisted in simple, programmable, remote-controlled switchboards with separate monitoring for each apparatus, for instance those for programming boilers and controlling pumps. Spurred by the energy crisis in the 1970s, these various installations were gradually connected together into networks so as to form complex automated systems. The aim was to save energy and reduce management costs. For example, supervision systems installed in lifts recorded and optimized responses to the calls on the various floors.
Developments in electronic technology in the 1980s further increased the possibilities offered by centralized control systems. Different automated systems were connected to each other, from the energy management systems to those for vertical transport, security, or fire preven-

tion. The objective was to optimize the systems and services provided in the building. In the 1990s, all the networks were integrated that carried digital information (computer data, telephone signals, and signals for environmental control and security). Networks which used to be installed independently are now connected according to the particular usage requirements. In the year 2000, the phenomenon of global networks such as the Internet and the possibility of connecting through a PC are separating the networks' functions from their infrastructure. At the same time, new digital technologies have opened the way for virtual configurations previously unthinkable.

NEW HIGH-TECH TERMS IN BUILDING CONSTRUCTION

Building area
- *building automation / b.a.*
- *smart building*
- *computer automated building / c.a.b.*
- *computer integrated building / c.i.b.*
- *immeuble precable*

Services area
- *shared tenant services*
- *multitenant communication building*
- *technology enhanced real estate*
- *immotique*

Home area
- *home systems /hos*
- *home automation /h.a.*
- *electronic home*
- *smart home*
- *home electronic system*
- *domotique*
- *Internet Home*

Though only in a general way and at times a bit ambiguous, it is interesting to note how these terms reflect each country's cultural and behavioral standards. For instance, in the US, where people's behavior tends to be pragmatic, i.e. based on immediate experience and direct experiment, terms such as shared tenant services and technology enhanced real estate are used for the most part to indicate that equipping a building with advanced management and control systems makes sense only if the economic profit involved is somehow quantifiable. In other words, advanced technology is valid only if it helps sell a building or rent all its rooms. In Japan, however, other terms are used such as computer automated building and multitenant communication building. Using extremely rationalized procedures and strict logic, a building is considered as intelligent as its connection networks are extensive, thereby enabling maximum communication. Finally, in Europe the models are more complex. Since much attention is paid to the well-being of the individual, advanced building automation is interesting only from the viewpoint of user-comfort. A building is smart if it is a pleasant place to live and work.

Going from left to right, we observe the progress from simple supervision to control and finally to intelligent self-regulation managed by a complex installation, as automated systems gradually become more complex and refined. Practically speaking, the left column represents traditional buildings. There are separate networks, sensors like those used to determine internal and external temperatures, and activators like those which regulate the boiler. On the other hand, the column on the right represents buildings with integrated automated systems where installations interact. It is interesting to note how, over the course of time, firms active in the building automation market have gradually and inexorably moved away from their original position as producers of

DIFFERENCES BETWEEN BUILDING AND HOME AUTOMATION

	Building automation	Home automation
Building type	large multi-storey building	family home, terraced
Use	offices, hotels, shopping centers	living-space
Users	many people	family, single people
Type of activity	work	spare time, house work
Interface operators	professionals	family members
Location in building	false ceilings, floating floors	walls
Installation	prefixed and recurrent schemes	ad hoc

MAP OF BUILDING AUTOMATION			
Service	maintenance surveillance meter reading	programmed maintenance	computerized management
		remote-maintenance remote systems management	
Type of Product	remote reading		computer integrated building
	programmable meters and timers	integrated systems	
	alarms	self regulating environmental systems	
	sensors and activators		
Apparatus works	single purpose networks		integrated networks
	single installation		integrated system
		level of complexity	

high-tech instruments (both hardware and software). From this position, represented by the central and lower part of the chart, they have looked to become service providers, as represented in the upper section of the chart. These services, much in demand and very profitable, are very attractive for such firms since they are readily quantifiable in money terms. This is the natural progress from supplying a product or automated system to supplying a "turn key" building and all related services.

2. How Does an Integrated Building Function

The difference between buildings provided with advanced automation and normal ones such as those surrounding us lies in the presence in the former of a central computer which functions very much like a human brain. This computer is linked to an integrated network with a role roughly equivalent to that of the spine in the human nervous system. Branches extend to all parts of the building while sensors and activators lie at the system's periphery. The different systems and installations are connected inside the building by means of an integrated network. The entire building can be kept under control since the various installations, from the air-conditioning to the security system and computer systems, interact by means of the network which forms the building's "spinal cord". This general design is valid both for building and home automation.

Merely installing cables and a computer in a building is not enough to make it smart; the systems might not even be connected physically but by means of an external network like Internet. Ultimately, what matters is the manner in which the integration was designed in line with user needs. Thus, a building could be considered "technologically advanced" when all the functions it performs are integrated and managed by an automated system. Since modern technology offers the building designer a wide choice of installations, it is possible to offer very sophisticated, "tailor-made" solutions and services to suit customer requirements.

The technological systems usually operative in an advanced building can be divided in four areas: management, control, information and communication. The first two have to do particularly with system installations, while the latter tend to be grouped with building services. The more advanced the building, the more these functional areas are connected. Management systems are those concerned with energy sources and maintaining ideal air and temperature conditions within the building.

It is interesting to underline that in advanced buildings the average standard of climate comfort is tending to disappear in favor of more dynamic systems which enable more personalized control.

Climate is controlled for small zones and areas, and lighting levels or air circulation speed are directly regulated by individual users.

As regards vertical transport systems, innovative aspects concern the ability of each lift to carry out self-diagnosis in order to react to potential malfunctions in real time, as well as the introduction of an electronic memory which enables the lifts to compute and predict the amount of traffic to be expected on a certain floor so as to regulate their operation speed according to the day and hour.

Control systems are those involved with surveillance and the general security of things and people within the building. Here, more than in the other areas, intelligence is distributed to the periphery where new sensors are able to process and filter information collected from the environment. In this case, it is the peripheral intelligence that reacts immediately, by making an alarm go off for instance, while the central computer registers the event and eventually sets further security procedures in motion.

The most interesting aspect of information systems is their ability to manage large quantities of data concerning the activity which takes place inside the building both at individual and overall levels. Systems like electronic data processing enable operations to spread much further than the actual building, as when a bank has branches scattered here and there. Information from all the branches, including system data from other buildings, can thus be gathered in a single place. Of course, networks such as the Internet exponentially multiply operational possibilities. From a construction viewpoint, the integration of information systems inside a building requires, apart from the use of special technical flooring, above all specially equipped spaces (Heavy Duty Zones) where the most important pieces of equipment are concentrated and, for important applications, can produce a lot of heat and noise.

Communication systems enable long-distance transmission of voices, images, and information from one part of the building to another or from the building to the exterior. These represent the most advanced part of a building with the most interesting technological transformations and the introduction of new services. From a technical viewpoint, the general tendency is one of gradual miniaturization. An optical fiber cable avoids the use of bunches of various cables and better conveys more information.

Procedures for interfacing and installing communication systems are becoming ever more complex because of this tendency towards miniaturization.

2.1 In-depth Analysis

BUILDING SYSTEMS
AREA OF SYSTEM MANAGEMENT

Energy
- *HVAC (Heating, Ventilation and Air-Conditioning)*
- *lighting*
- *energy management*

Technical Systems
- *vertical transport*
- *public services (canteen, cafeteria, etc.)*

BUILDING SYSTEMS
AREA OF CONTROL AND SECURITY

Security
- *access control*
- *presence detectors*
- *anti-intrusion*
- *anti-burglary*
- *video surveillance*

Safety
- *fire prevention*
- *gas and water detection*
- *electric safety*
- *earthquake detection*
- *evacuation management*

BUILDING SYSTEMS
INFORMATION AREA

Office Automation
- *word-processing*

- *electronic filing*
- *e-mail*
- *desktop publishing*

Specific Services
- *CAD/CAM*
- *administration*
- *records*

Electronic Data Processing
- *network connection*
- *back-up and archives*

BUILDING SYSTEMS
COMMUNICATION AREA

Audio-phonic
- *word-processing*
- *telephones*
- *intercommunicators*
- *telex*
- *facsimile*
- *data-base access*
- *message distribution*
- *Internet*

Images
- *entrance videophone*
- *slow-scan video*
- *videoconferencing*
- *closed circuit video control*

CONTROL OF LIFT MANAGEMENT SYSTEMS

The principle aim of lift management systems is to minimize waiting time. This is done by combining control and analysis of the movements of two or more lifts, of the calls from the floors or from inside the lifts, and of the position of the lifts and the people waiting to be transported. Combining this data enables optimized use, for instance, in order to reduce the number of stops per lift or send a different lift from that which was called. Today, a few advanced models are already able to ''see'' with infrared sensors how many people are waiting on each floor and regulate operation speed accordingly. Since computerized management keeps records of all events, it is possible to determine what parts need preventive maintenance if they have already malfunctioned on more occasions than is statistically acceptable. Again regarding self-diagnosis, one of the main problems concerning automated vertical transport is the shifting out of phase of the information gathering sensors located above and below each stop. With a traditional system, a technician must be called whereas an advanced system is able, once the lift has reached the ground floor, to perform the re-phasing operation automatically because of an initializing procedure managed by the central computer. Similarly, sophisticated lifts can perform average self-maintenance and repair operations so that a technician is only required in cases of serious malfunction. In Italy, perhaps the most interesting application stems from recent safety regulations (UNI EN 81-1) for personal and freight lifts which require a system of vocal communication to be connected to the cabin from a central operations room as a precaution in case of breakdown. Since the presence of this transmission line is now a legal requirement, it might also be used to send data to a control center thereby facilitating self-diagnosis and lift management.

Commerzbank in Frankfurt.

3. Cablings

Installations in the different parts of a building are interconnected by means of cable networks which convey all the data and control signals and are combined with power cables. Accordingly, the network of cables is the most important component in an advanced technology building, as well as the component which presents the most potential for further development. Pre-installing such a network provides the building with a basic system engineering structure that lends itself to any application. Any kind of device can be peripherally connected to it, from phone terminals to temperature sensors. It is no longer necessary to install new cablings for each new application as in traditional "peer to peer" buildings. The central computer only needs to be set accordingly.

Though a body of regulations exists for cablings, there are as yet no internationally recognized standards. Though many commissions are active in this regard, the pressure from producers is so strong that only very general criteria are defined to try and content all of them. The development of the market has nevertheless led to certain "de facto" standards put forward by the most important producers in the field of information and telecommunications technologies.

No matter what network is installed, the basic configuration always follows the same design. Special ducts carry the integrated network through the entire height of the building thus forming a kind of nervous system. On each floor, the cable network is first of all connected to the lift system. Then the cables branch off, usually beneath technical hollow floors, to reach all the individual application points. Starting with this basic design, all possible configurations can be created to meet specific requirements and optimize the cost/performance relationship.

There are three ways of physically distributing a network within a building: peer to peer, hub, and bus. These can also be arranged in mixed configurations. Moving past the classic "peer to peer", similar to that found in most dwellings, unable to carry much information since it is limited to a departure and an arrival point, so-called "hub" cable networks open from a central point outwards and allow many different configurations. Their importance has been immense. Ethernet, one example of "hub" distribution, was developed by Xerox. It uses a coaxial cable with double shielding which is usually yellow. Decnet was a later derivation under the Digital brand and offered new products ranging from local repeaters to a connection bridge which enabled various cable sections to be connected. The cable, called "thin wire" and usually blue-colored, is thinner and more flexible than Ethernet cable. It offers minimal encumbrance and a better bend radius. IBM's Token Ring is a similar network but configured like a ring. Certain data (tokens) give permission to use the network for a predetermined period and circulate along the ring until arriving at the intended work station(s).

Bus configurations, consisting of a spine with lateral branches, were conceived for simpler applications. However, they are now also used for more complex projects. The basic idea of Bus is very simple: sensors and switches are connected to a low-tension line, usually made of one or two duplex telephone cables, and coupled with the power installation. Since any device, no matter what its function, can be connected to it, the building can be fully cabled before the system is planned in detail. Thanks to microprocessors in each apparatus, it is then enough to define codes for the individual devices to configure the entire system. This avoids problems linked with installations behind finished walls, such as having to remove wires when anything must be changed.

Producers have put forward many effective standard systems, among them the X-10 whose 20 years on the market have made its importance almost legendary, the North American CEBus, and the revolutionary Echelon which surpasses the concept of Bus, the Japanese HBS and THBS, and the European EIB, BatiBUS, and EHS.

With the appearance of the Internet and Intranets, even the world of communication networks within buildings is changing and firms in the field of automation are developing solutions based on new technology. For example, the Echelon Ipi.Lon1000 which makes it possible for any kind of application, apparatus and user to be connected to the Internet and incorporates a Web Server by which Web pages can be created to surf the Web and exchange information in real time with connected devices. In fact, new networks are more and more able to create what is called a vast "distributed environment," i.e. to supply anyone with data, no matter where they are, as long as they can connect to the Internet, for example via a simple desktop or laptop PC. Furthermore, management resources and applications are made available to the Web, thus solving the problem of specific software as soon as it is necessary.

3.1 In-depth Analysis

ADVANCED CABLING SYSTEMS AND MARKET REACTION

Cabling inside a building has become the most important part of system engineering and that aspect which, more than any other, makes it possible to optimize the use of all the installations equipping a building. It enables connected devices and systems to interact using a common language. It is important to shed some light on a few concepts and say a few words about the present state of the market. There has been much talk about buildings with advanced systems integration such as Computer Integrated Buildings dependent on so-called "cabling systems," multi-service networks integrating voice and numerical data, such as Ethernet, Token Ring, etc. This image of the integrated building was very attractive for systems producers, especially those of computer systems along with telecommunications, since it shifts to the building construction field a trend now established in many other production areas: the ever growing role played by services. Because of the complexity involved in design and installation, the cost of advanced integration is very high which means its use is optimal in buildings with very large floor spaces and a single customer. Accordingly, use of large multi-service networks will probably be restricted to service sector areas and buildings more than others as testimony to the prestige of high-tech systems. However, market development, in response to small offices, stores and homes, will favor less complex projects. The most common networks will be simpler to install and operated more or less in connection with the Internet.
In this regard, the Japanese firm NTT offers an interesting example of marketing strategy. In the three Shinagawa towers, the firm's headquar-

ters built in Tokyo, the building installations are integrally interconnected by means of a new kind of cabling network. Moreover, innovative technology and new office services are being tried. The same company built another office building in Tokyo, though not as their headquarters but as rental office space. The entire system was conceived with this function in mind, though other basic services are provided at a high level of sophistication. In this building the centralized system was a simplified basic structure and each separate zone is equipped with its own management and control system in terms of automation, lighting, and telecommunications structures. Another NTT building in Tokyo with rental spaces has very particular requirements since it accommodates both offices and research laboratories. In this case, the building is characterized by a centralized fiber optic network for data and improves the automation systems already present for HVAC, lighting (lights are automatically turned off when no-one is present) and all security installations. Office automation systems make possible computerized programming of the workload, automatic transfer of documents, and offer support for planning and organizing conferences and meetings.

No matter how complex they are, networks for simplified data transmission must guarantee secure control and technical management of the building and must be more or less compatible with systems for the transmission of images.

STANDARD MODELS FOR BUS NETWORKS

X-10

Developed in 1976 by the American firm Pico Electronics and mainly used in homes and small shops. Its great strength was and still is the great flexibility it makes possible and the fact that it uses the common electric wire as a means of transmission. In practice, a small module able to send and receive impulse modulations is placed between the plug and the controlled device. The system has a free configuration and can be designed over successive periods of time. The modules can even be purchased in department stores. The system is able to cover more than 250 points and can be directly managed with special simplified programs from a small PC.

CEBus

First developments on this started in 1982 when the American firm Electronic Industries Alliance became aware of the fact that computers

*were becoming an integral part of everyday products. It was complet-
ed in 1992 and became the North American standard model with the
EIA/ANSI-600 specification. One of the biggest problems Americans
have had to face concerns the high number of remote controls used
sometimes in the same room and at times with the same transmission
codes. Accordingly, CEBus transmits data predominantly through the
network of electric wires. In the most extensive configurations, up to 4
billion combinations are possible.*

*The later model Plug'nPlay, distributed with the EIA-721 specification,
was based on this, without however the specifics related to means of
free transmission. Intellon Corporation and Domosys are the principal
producers of modules for Plug'nPlay.*

Smart House

*The basic idea stemmed from the observation that each house contains a
large amount of different cables, each with a specific function and each
installed by separate technicians without any interconnection. Thus, a
consortium of American producers conceived of integrating all the various installations by using a single form of socket into which any device
could be plugged, from the toaster to the telephone to the washing
machine, so that any kind of communication and any service could be
provided in every part of the house. Since everything is managed by a
central computer which forms an integral part of the system, very high
safety levels are obtained. For example, if the computer does not recognize the apparatus requiring energy as suited for a certain plug, it does
not supply the energy. It is no longer possible for a child to be electrocut-
ed since energy will not be provided to a device which is not functioning
properly or losing current. When this is the case, the computer immediately becomes aware of it and shuts down the energy flow. Thus it supplies electricity according to the device which is connected to a plug.
Because of its configuration, Smart House is a closed system with which
producers can only communicate by using regular usage licenses.*

LonWorks

*Better known as Echelon (nothing to do with the American global spying
system) from the name of the firm which developed the product and
which numbers such names as Motorola and Detroit Edison among its
founders. Unlike other bus systems, it constitutes a proper technological
control platform which can cover up to 32,000 different points. It is based*

on intelligent nodes provided with processors (NEURON CHIPS), control and detection devices, and transponders which interact with their surroundings and between each other using all possible means of communication including the Internet and similar networks like Lan or Wan based on similar widespread transmission protocols. It is also the only bus system which can be installed not only in buildings but also in cars, assembly lines, etc. LonWorks is used both in the new line of intelligent appliances from Merloni, and the new digital meters used by ENEL, the Italian Electric Company, which will make different kinds of rates possible and enable people to control their home appliances by phone.

HBS

The first model to be officially recognized by the Japanese Ministry of International Trade and Industry (MITI) in 1988. Both NTT and the Kansai center for the development of industrial electronics collaborated on this project. Essentially, it is an audio-video and telephone transmission system which enables special installations to interact not only with each other but also with external networks. An extended model called Super HBS was developed to operate in both buildings as well as city blocks. It provides a main communications backbone to single buildings and complexes of about 300 apartments (typically found in Japan).

NEC THBS

This symbol appeared when the Japanese firm NEC developed its "Total Home Bus System". It was conceived as an integral part of the home electric system to be connected with HBS by means of a series of interfaces. The main element is a device called a "telecontroller" which functions as a link with the external telephone line. It is connected to the central computer of the apartment which usually has a built-in phone. Apart from this, there are many points which can be activated by infrared remote-control. Above all, the system controls internal communication (video entrance phone), lighting, climate control and electronic doors which open with a digital code instead of a key. The system can be activated both from inside as well as outside.

EIB

Initially developed for business purposes in 1990 by the European Bus Association (EIBA). It is the standard model for Siemens products. Essentially, it consists of a special duplex cable which enables the serial *transmission of data using a specific communication protocol. It can also use electric cables and radio frequencies and, in special cases, be connected to the Internet. Each appliance controlled by this system contains a special electronic card used to receive commands. Thanks to special relays, the system enables more than 61,000 devices to be connected. For the most part, the devices connected to the system are electrical and control technical installations and surveillance.*

BatiBUS

A communication system developed in 1988 by Merlin Gerin and EDF for the control and management of technical installations, lighting, and security. It was the first European model to be put on sale. It uses a twisted duplex telephone cable whose length varies according to the requirements. In the BatiBUS system, intelligence is distributed since each component contains a microprocessor which independently processes information and manages data communication. A single cable supports the entire system. The network can serve up to 256 different points and can be connected to local or area networks.

4. The Future of Design

While the building contractors and designer play a much more important role than the system engineer in traditional building projects, in high-tech buildings the most important figures are those on whose intervention the good operation of the building depends since this is the ultimate criterion of the success of the entire project. These new figures are the system installers and integrators, specialized firms whose task it is to arrange for systems and devices to interact. The suppliers have always assumed this role as it goes hand in hand with the sale of devices and the installation of networks. Thus, highly automated buildings and the technical developments they represent are profoundly transforming the world of building construction and especially the relationship and importance of the "classic" construction operation with respect to that part considered "non-construction" such as systems installations. The importance of building contractors and designers has receded while very specialized operators have come to the fore, employees of multinational firms used to business logic and with managerial capabilities usually lacking in designers. It seldom occurs that a designer, or design-team, or even building contractors with system engineering know-how are able to plan the construction of a building with advanced integration.

Accordingly, strategies which take recent changes into account must be developed to approach such projects. The designer must come to grips with the workings of high-tech systems and collaborate with system integrators from the very beginning of a project. Being aware of the potential presented by new networks and more and more intelligent installations, the system integrator can design projects whose layout and structure aptly reflect and exploit new technologies, concentrating on the quality and livability of the spaces. The possibilities are endless and success will depend not so much on training as on a willingness to invest in the future and individual creativity. The most fertile areas at present are those of the outer shell of the building and the division of the spaces. The use of automated glass roofs which connect several buildings to each other, though already widely uti-

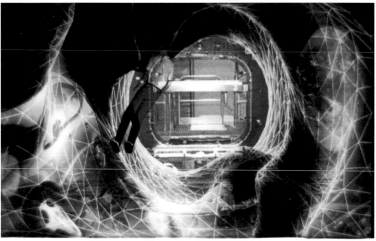

Facing page: www.malibu.video.house.pch. *This page: above left and below,* Trans-Port 2001, *an Internet-linked installation which creates virtual surroundings; right, MITs "Pfinder" interactive walls.*

Trans-Port 2001, *Internet-linked installation which creates virtual surroundings.*

lized, could be further developed using new and more efficient formal solutions, exploiting "double skin" structures such as the one adopted for the Frankfurt Commerzbank, or "triple skin" ones like the Helicon in London or the Green building, structures which enable natural ventilation during a good part of the year and also new internal mechanisms. Similarly, the use of the new "cold ceiling" system opens new vistas for designing a building's volumes since the external perimeter is freed from the distribution networks and fan-coil furnishings.

Considering how networks such as the Internet are separating a building's functions from its infrastructure and the infinity of new ways of accessing such networks apart from the PC, it seems probable that design constraints which stem from system engineering requirements will gradually disappear. Even now, further and unlimited settings are conceivable which explore not only the physical spaces design has already taken into account but also the virtual spaces which new technology has made possible. Amongst other possible examples, mention could be made of the interactive walls developed at the MIT Media Lab which are able to recognize a person and their mood by using biological models and energy maps of the human body; as well as in Michael Jantzen's Californian house whose façade is covered with video screens relaying images from the beach behind it and whose windows work both as openings and as transmission screens. The project by Kas Oosterhuis is even more forward-looking. He is designing a virtual environment with flexible membranes able to assume various volumetric shapes and where inner grids project images and texts which come from the Web.

www.malibu.video.house.pch.

NMB BANK IN AMSTERDAM

The headquarters of the Nederlandsche Middenstandbank (NMB Bank), situated in the commercial area southeast of Amsterdam, is a very original and evocative building and offers a contrast to rather anonymous, standard office blocks. The building represents an attempt to create a "different" workplace where attention toward the needs of the human personality is combined with efficiency and flexibility. The principal aim was to obtain the best possible balance between, on the one hand, technical and organizational aspects, and on the other, the needs of the employees actually at work in the offices: bankers, if considered from a professional point of view, along with personnel even seen as individuals. This represents quite a challenge given the high number of employees, more than 2500, working together in this one complex.

The complex was designed by Dutch architect Ton Alberts with a concept based on the idea that an office is a little like a third skin. The first skin is our epidermis, the second is our clothes, and the third is the building where we work. If the latter is only barely adequate, people experience discomfort and are eager to leave their workplace so we often hear phrases such as "Thank god, only a few minutes before we can go home!" Such a relationship with work quickly becomes extremely counter productive while a pleasant work environment is a place where people seem not to be working but simply doing things. The chosen design was to avoid the usual tall monolithic building and erect ten blocks, 3 to 6 stories high each, connected to each other by pedestrian galleries to create a ground plan shaped like a large irregular S. Viewed from outside, the most striking aspect is the walls inclined at various

angles. Apart from their unusual aspect which, among other things, necessitated the use of special bricks to cover the slanting surfaces, this particular characteristic meets precise needs in terms of sound absorption and energy savings. Not only does the inclination of the walls deflect traffic noise towards

Bannister with small waterfall.

the sky, just like those used on busy motorways, but it also enables better reception of the sun's rays since the angle of inclination of the walls was determined to obtain an optimal angle toward the sun. The irregular façade neutralizes wind stress, further reducing energy consumption. Thanks to the T-shaped plan of each block and the links connecting one block to the other, giving an overall S-shape, sheltered green spaces have been created, planted with high trees which contribute to summer cooling and better protect the building from winter winds.

The concern with energy savings determined not only the blocks' exterior shape but also their internal structure. The roof has broad pentagonal glass surfaces which absorb and store solar energy and constitute an integral part of the heating and ventilation system. The latter, along with all system networks, runs down a special central well in each building. On cold sunny winter days, a good part of the energy requirements are covered by solar heat and the heat generated inside by people and equipment which is extracted from the ventilation system by exchangers. Thanks to all these devices, the NMB headquarters building consumes relatively little energy per square meter. In all, it consumes 111 kwh/square meter a year, or 96 kwh/square meter without taking into account dispersion to neighboring houses and shops. These figures are far below the average in northern countries for more or less traditional buildings which reaches 700 kwh/square meter in banks and 500 kwh/square meter in public offices.

For purposes of energy saving, each employee is to regulate the temperature in his own workplace. Windows can be directly opened and closed, and shutters and blinds directly adjusted, though they automatically adjust themselves when solar rays on the façade goes above a certain level. In any case, the automatic ventilation system maintains a constant temperature throughout the entire complex.

Even though many lifts connect the floors to one another, many stairs link one hallway to another and facilitate rapid movement on foot, especially

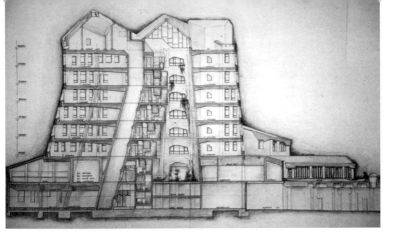

Above: NMB Amsterdam, section. Below: stairs between floors.

from one floor to the next. The staircases are abundantly decorated with tropical plants and are used as a "humid lung" within the system of energy exchanges. Paintings, art objects and marble or bronze sculptures are arranged around all common spaces in combination with fountains and ponds. Water is used to provide continuity and is found throughout the complex: in the Japanese gardens between the blocks, in roof gardens where sculpted art forms surround it or let it flow, even in a brass handrail where water runs down two floors to complete its course as a small waterfall in one of the main entrance halls.

SAS HEADQUARTERS IN STOCKHOLM

The new headquarters of the Scandinavian airline SAS are located in Froesundavik, just north of Stockholm, midway between the Swedish capital and Arlanda International Airport. The new corporate complex brings together departments which were previously scattered in 30 different offices.

To compensate for the size and extension of the complex and promote human exchanges, the complex was designed as a kind of village whose reference point is a central avenue, which the various buildings flank, and with a protective transparent shell which covers all open spaces.

This central avenue constitutes an unusual reference point for a corporate complex. Shops, a restaurant, two bars and a reading room all meet around this axis. The service areas are easily accessible from it, as well as the gym, covered swimming pool and other facilities provided for the employees to spend any spare hours. The buildings are made with a steel skeleton structure

and prefabricated light ferroconcrete panels. All outwards facing fenestration (facades and roofs) is double-glazed while windows facing inwards are single-glazed with glass covered by a slightly reflecting film.

The complex's main feature is the intelligent management of the HVAC system which uses the aquiferous basin on which the building is erected as a heat reservoir.

Water-bearing basins are nothing but large subterranean tanks from which water can be drawn for large and small cities. In Sweden, these are usually of morainic sand and rock deposited after the passage of glaciers at the end of the last ice age and surrounded by more impermeable rocks. By simply allowing water at the requisite temperature to filter through them, such basins may be used to store thermal energy both on a day-to-day basis and

from one season to the next. Apt use of pumps enables hot water to be stored during the summer which can then be used for heating in the winter. On account of their different density, two distinct masses of water are formed underground, one hot and the other cold. There is no need to renew water supplies since all the system does is transfer it from one place to another. The aquiferous volume at Froesundavik reaches about 1.5 million square meters while the basin extends at various points from 100 to 200 meters in width and 15 to 30 meters in depth. Since natural underground water infiltration and downflow are extremely reduced, the location was ideally suited to be used as an energy bank.

At the moment, 5 wells enable water circulation around the new SAS complex. Two of these are used in the winter to draw water from the areas where hot water is collected. From the wells, the water goes through exchangers, which extract the heat for the building, and then returns to the underground reservoir through one of the three wells which connect with

the colder water areas. In the summer the procedure is reversed.

The system provides cooling water for the equipment in all offices, hot water for the sanitary fixtures, pre-heating and complete cooling of the air used in the air-conditioning system for the central area around which the other buildings gravitate.

As regards the offices themselves, they function according to the "box within a box" principle. The air-conditioning system and perfect air-tightness of the "external box," i.e. of the whole complex, prevents the need for further heating during work-hours since the heat generated by the computers is sufficient. Only during the coldest winter nights is an auxiliary heating system automatically turned on with electric panels to avoid excessive energy loss. A small unit around which cold fluid circulates provides cooling during the

summer months. There are neither ventilators nor ventilation ducts. While heating panels and cooling units are placed above the windows, the openings for air circulation are positioned on the intrados of the internal wall. Through these openings, air from the main avenue passes through a conduit where it is warmed or cooled according to the season. Office temperature is totally independent from the ventilation system and can be regulated separately by each employee. However, levels above or below average are centrally recorded since they require slight modifications in the system's operation.

The complex of sensors, switches, pumps and thermal exchangers is completely automatic and performs the task of combining the heating-cooling resources available from the water-bearing basin with those provided by the large glass roof. During the summer months, cold underground water is passed through the exchangers thus providing cold air for the system which "recharges" the basin by using heat extracted from the hot air which accumulates in the central gallery of the complex. When cooling is not sufficient, the system opens the skylights to create a slight draught inside the complex. On the contrary, during the winter, the system both pre-heats the air by using the subterranean hot water and controls the levels of heat extracted from the air in the offices along with heat provided by the greenhouse effect of the roof.

Full operation of the system in the last few years has made it possible to collect general management data. Usually, heat is stored between May and September, with high-points in June and July and monthly averages of about 300 megawatt hours. This heat is then used between November and March. In April and October the direction of the energy exchange varies from day to day. Cooling in the offices not only takes place in the summer but also in winter with overall requirements averaging 200 megawatt hours per month. Most cooling in winter is due to the direct action of employees too impatient to wait for the system to react to temperature changes. A series of previsions to be added to the system are being studied in order to avoid such problems in the future. Finally, heating requirements have proven even more seasonal than those for cooling, with peaks in November and February and averages of around 150 megawatt hours per month.

A further indicator can be drawn from energy requirements per square meter. Out of a total of 64,000 square meters, 55 kilowatt hours per square meter are used for heating, and 50 kilowatt hours per square meter for cooling. These figures exclude direct electric heating turned on by the individual employees.

Fair Tower in Frankfurt

The Fair Tower in Frankfurt is one of the best examples of a building whose form was conceived to exploit as much as possible the advantages offered by technological systems.

The building's heating and air-conditioning systems are able to automatically adapt themselves to previously computed data. Not only does the system continuously manage the building's thermal requirements according to user needs but it also draws on available data to optimize its functions thanks to a smart system that calculates new use parameters each time.

More specifically, the offices in the tower are served by two automated air-circulation systems with 74,000 cm/hour air capacity and two other systems with 98,000 cm/hour. Heating and cooling are carried out with high-pressure steam provided by Frankfurt's municipal network and used either as a primary fluid or through thermal exchangers. External air is conveyed to the central system, filtered, and finally heated or cooled by the exchangers according to need. By means of a radial ventilator, it is then sent to the various floors of the building where a flux regulator controls its intake according to air-pressure. These installations are connected to various automated subsystems which control air supply, interrupting it if need be, for instance in case of fire. Office temperature is regulated by more than 2000 convectors placed along the external perimeter beneath the windows. Air-conditioning

cycles are controlled by 35 special sub-systems each able to operate autonomously. Data and operations are gathered and transmitted to the control point in the central security offices situated in the tower basement where they are shown on color monitors. Malfunctions are signaled at the central office and the entry desk where instructions regarding procedures to follow are printed out thus providing technicians with the necessary information.

System integration, however, is most effective in the area of security, particularly that of fire prevention. The building can accommodate 4000 people on 70 stories and 2 basement levels. Should a fire break out, not only would all the people be evacuated in the least amount of time but firefighters would be able to act with hydrants even at considerable heights. This is a sensitive issue in Frankfurt where, over the last 20 years, various fires in office high-rise buildings have demonstrated that firemen are often unable to act effectively in these situations.

In the Messeturm, it has been repeatedly proven that firefighters are able to reach even the highest floor in less than 3 minutes. If need be, they can evacuate everyone at most in 15 minutes. As soon as a fire breaks out, a sophisticated alarm system goes off. It is connected to a very dense network of smoke detectors and sprinklers. There are more than 900 smoke detectors installed on office ceilings, in thermal systems, and in lift shafts, as well as more than 350 push-button alarms on staircase landings near extinguishers. The whole system is supervised by a central computer which makes optimal use of its technical possibilities. This system surpasses Frankfurt's strictest building standards and is considered an example worldwide.

The principal criterion of the fire prevention system is to ensure the greatest possible safety for the people inside the building. For this reason each floor is divided into 4 security zones equipped with fireproof doors, each with its own stair access. In the event of danger, a secure evacuation area can be reached in less than 10 meters..

When the alarm goes off, the building responds automatically and simultaneously performs a series of complex procedures lasting just a few seconds. First of all, on the affected floor and its staircase, and those immediately above and below, both acoustic and visual alarm signals are set off. The latter were installed in great numbers especially in areas where high sound levels are expected. Then, on the same three floors, the ventilation system is cut off while air-pressurization systems are activated. On the affected floor, only small air outlets remain open connected through special valves to the smoke-elimination system. The fireproof doors close while the security exits

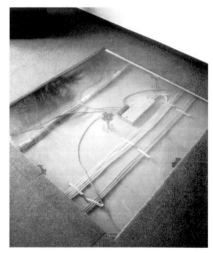

Above: control room. Below: left, specially equipped ceiling-panels; right, specially equipped flooring.

open. When people open doors to evacuate, the smoke cannot go from one zone to the next because the system highly increases air-pressure in all four security areas. All the lifts are sent to the ground floor and special panels are activated where the ground plan of each floor can be seen with light signals enabling firefighters to immediately identify the point where they must act. As soon as the sprinkler system is activated, the central system

automatically sends an alarm signal to the fire brigade. Two special smoke-proof lifts bring the firefighters to the affected floor. There are 4 fire hydrants on each floor, as well as manual extinguishers kept in special closets.

The Messeturm's elaborate installations are completed by a sophisticated anti-intrusion security system. Movement sensors and magnetic doors control the entire building. Special badges are use to gain access to particular areas. Whenever no presence is detected within an area, all doors are automatically locked. The central control area is situated in the main entrance hall on the ground floor, where turnstiles automatically allow badge-wearers to go through to the main hall and its 24 lifts. On the tower's north side there are also turnstiles for people in wheelchairs. An important point concerns visitor access cards. Once they have reached the area for which they are cleared the cards switch to exit mode thus preventing further access should the cards be handed to non-authorized people.

ABB HEADQUARTERS AT SESTO SAN GIOVANNI, MILAN

The ABB headquarters is the first building in Italy equipped with the new "cold ceiling" technology. This climate controlled ceiling consists of a false ceiling covered with aluminum panels in which a copper tube conveys hot or cold air, depending on the outside temperature.
In order to avoid acoustic reverberation, more than 50% of the aluminum surface is perforated. The result is a kind of grid where the only filled points are those with lights. Thus, the entire false-ceiling functions both as a light-radiating ceiling and a temperature-control system.
The system offers the following advantages:
– it provides a uniform heat-absorbing surface;
– since air movement is reduced there are fewer draughts, though air circulation and ventilation remain adequate;
– it can be adjusted to meet the requirements of each different zone;
– it can be adapted to even the most unusual architectural design;
– it is controlled by a single on-off valve which is the only part requiring maintenance;
– it uses water at 16 to 17 degrees Celsius which it heats at lower cost than traditional systems. It makes use of natural resources such as fresh water from lakes and rivers where this is permitted;
– it increases energy savings from 15% to 30%;
– it is silent and the low volume of air exchange can easily be set at an established sound level;
– the system recharges itself since the ceiling absorbs any excess heat radiated from the room surfaces;
– in areas that generate higher levels of heat, such as machine or computer rooms, an additional system is provided to manage the excess heat.

Giancarlo Marzorati, architect and designer of the project
"Having overcome our initial perplexity, the use of such a sophisticated system greatly stimulated the design, especially because of the space it created. Above all, since the false ceilings eliminated the usual rigid constraints, it was possible to design an S-shaped building, i.e. with a wave-shaped façade instead of classical volumes with flat surfaces. The ceilings can be adapted to any spatial configuration and the only problem was finding a system to join the ceilings to the walls which retained the highest possible flexibility for possible new systems of environmental control. At the ABB

headquarters, the lines of the false ceiling grid coincide with the central axes of the façade windows. To be certain of adequate lighting, no matter how large the rooms, light sources in the false ceiling are positioned in the centers of the grid. There are no ventilation convectors which usually run along the perimeter of the building forming a cumbersome ring around each floor. Since the windows reach down to the floors, it is possible to obtain special effects with light on the flooring. These esthetic and expressive characteristics go hand in hand with advanced technology and it is likely future office buildings will have similar features.

The main entrance faces the new street and leads you into a large hall

Cold ceiling.

bathed in sunlight. Essentially, this is a three-story high atrium with a network of girders supporting the ceiling. The nearby conference room as well as the area with the information and reception points and the space for visitors on the ground floor required similar structures.

Because of the amount of natural light flooding the interior, the entrance hall was designed as a kind of oasis with a pond and high semitropical trees instead of the classical marble floors and facings. With the canteen facing this space, the aim was to create a pleasant and comfortable environment where people have the sensation of being in another place."

HOUSE OF THE FUTURE AT ROSMALEN (HOLLAND)

Built by Libema Vrijtijdsparke, a company that constructs theme parks, Intervam, building contractors, and Chriet Titulaer, the designer of the project, the House of the Future *is situated in the south of Holland inside an amusement park. Apart from the* House of the Future *per se, the complex comprises another building which houses an auditorium, an information point, and high-tech gift shop.*

In order to minimize construction time and allow successive modifications to the walls and internal partitions, the house was built with a steel skeleton. Ferroconcrete prefabricated walls and parts are reinforced with pre-stressed fibers instead of the usual steel bars. Concern for environmental issues led to the use of a mixture made with recycled cement for the foundations. Similarly, the garden walls contain various types of recycled products, while internal wall facings are made of recycled paper and plaster obtained from waste from the desulphurisation of industrial products. Concern for the environment is one of the principal starting points of this project.

The house is equipped with five different containers for the collection and elimination of refuse: for colored glass, non-colored glass, batteries, paper, and organic matter (fruit and vegetables). Refuse from the last category is directly thrown out from the kitchen into a special container outside where it dries thanks to heat obtained from the house's ventilation system. The resulting material is then ground up.

Rainwater is collected in a large membrane stretching above the roof, also

used to provide shade. The water collected in this way can be used for the toilets and garden so as to reduce the use of drinkable water.

Six panels fitted with solar cells which provide electricity for the security system are set above the southern façade. A solar boiler which heats water for the bathroom is located on the roof.

Chriet Titulaer, Dutch astrophysicist, creator of the *House of the Future* in Rosmalen.

"On 20 June 1989, when Dutch minister Ed Nijpels inaugurated the House of the Future *(Huis van de Toekomst) at Rosmalen, an old childhood dream of mine became reality. I'd literally spent the entire week before inside the building 24 hours a day. I even slept there in order to be able to supervise all the work shifts following one another without interruption in order to meet the deadline. Though there were more than 250 workers on the site, we couldn't be certain we'd finish in time. We completed the lane, of recycled asphalt, leading to the entrance only the night before the inauguration. Apart from the short construction time, less than a year, the difficulty consisted in this project being for a house of the future, that is to say, involving experimentation with new building and electronic technologies as well as innovative applications of traditional materials. It must be noted that at least a dozen years of experimentation through direct application were required, making gradual changes and additions, to allow possible improvements in the domestic sector the time to mature. When I started designing the project, my main aim was to have an instrument which would enable us to consider future homes. I was concerned not so much with applying and displaying particular technological developments but rather with attempting to understand, based on current changes in society, what changes home life will undergo in the quickly approaching third millennium.*

Though future homes will most probably not resemble the house we have just completed, it is important to provide indications as to the main direction changes may take, and above all to stimulate the imagination of visitors as to what the future will bring, especially since the speed with which transforma-

tions follow each other is always increasing. In order to determine the features and shape of the structure of the house at Rosmalen, we carried out a comprehensive examination of the way modern society is changing. From this survey we derived a series of concepts we took as our starting points. Here is a summary of the principal ideas that have influenced the design of the house.

- As a result of ever increasing human activity, the environment is filled with cumbersome and polluting objects. It is important to protect the environment so that our legacy to posterity is not a destroyed environment.

- Life expectancy is always increasing and this tendency leads us to believe that people between 55 and 75 will be more and more active.

- Along with increased activity in old age, individual expectations will grow and domestic spaces must inevitably change especially as regards health and safety, communication with those around us, and the ease of making purchases.

- Contemporary life is marked by a constant rise in standards of living implying long-term consequences for domestic habits and spaces. The bathroom, for instance, apart from being necessary for personal hygiene, is used more and more as a space for relaxation.

- Permanent education, pursued throughout an individual's life, now appears as a sine qua non in order to be able to grow in modern society.

- The three most important socio-cultural developments in recent years appear to be first of all the expression and coexistence of diverse requirements, i.e. the polymorphism of our needs, then an ever-increasing discovery of individual values, and finally a process tending towards autonomy and emancipation.

House of the Future *at Rosmalen, telecommunication desk.*

- Flexibility and adaptability are therefore crucial for the building of the future.
Inside the house a communication system connects all the rooms so that every electric device can be automated and controlled at a distance. The house is also provided with a telecommunications network compatible with new standards. The result is a complex integration of systems and devices, all managed by a central computer. There is another computer in the work area which serves work purposes without any interface with domestic functions.
An integrated security system emits a sound in the case of danger and sends an alarm by telephone to security services. It also manages the main entrances by means of magnetic cards.
One of the aspects that most strikes visitors is the absence of light switches on the walls since the light is turned on and off automatically by means of small infrared sensors distributed on the ceiling. Behind each sensor is a switch connected to the central computer.
Another peculiarity is the circular bathroom entirely enclosed by transparent glass walls. This design is expressly provocative in order to draw the visitor's

attention to the fact that the house of the future will have to take account not only technological changes but also changes in social habits. The bathroom ceiling, glazed as well, is divided into segments which open like flower petals. These are controlled by a hydraulic system activated by vocal sensors. The bath is equipped with a system for hydro-massage and thermal bath. The glass wall and mirror surfaces are covered with a special coating which prevents condensation.

The House of the Future *at Rosmalen contains so many innovations it is impossible to list them all. In terms of robotic elements there is, for instance, a device able to read the paper with an artificial voice, as well as a lamp*

Specially equipped kitchen-space.

which obeys vocal commands and assumes six different positions without being equipped with a motor. This is made possible by a new technology which enables a compound of titanium and nickel to 'remember' the specific form it was made to assume at a particular temperature. By raising or lowering the temperature of an internal wire with electric resistors, the arm carrying the light-source moves to prefixed positions such as, for instance, 'to the left', 'to the right', 'up', ''down', etc. The development of this product as well as the facilitated use it offers the disabled clearly demonstrate how it is not only possible but important to create different ways of using everyday objects. Such improvements will result in a better quality of everyday life.''

TRON HOUSE IN TOKYO

The Tron house is made entirely from glass and wood. It was conceived as a one-family home for a nucleus of three people and covers an overall surface of 370 square meters. The whole southern façade consists of a glass wall with 100 glass panels. These, like all the windows in the house, open and close automatically thanks to a computerized system which monitors air conditions inside and outside. Data regarding temperature, humidity, barometric pressure, wind direction and velocity, rainfall, and lighting levels are collected for this purpose.

The house contains 1000 computerized circuits, connected in networks, and divided into 400 sub-systems. Thus, there are separate networks for sensors, the telephone, video data, the sound system, lighting, security, air-conditioning, etc.

Apart from internal and external communication, the telephone network enables interaction between the various computers, while the video system, in addition to sending images all over the house, controls entry into the house and a few specific areas and monitors all the sub-systems.

The sound-system makes it possible to listen to the radio and stereo from any point in the house. Because of a digital processor, working in connection with various other systems such as lighting, it also controls the sound volume to create various special environments, for instance, for a reception or a moment of quiet reflection.

In the kitchen, the different stages and procedures in cooking a particular dish can be recorded, including details for cooking temperature and spices,

in order to be able to prepare the exact same dish again and again. A multi-media system called BTron functions as an assistant and guides you step by step through the recipes of famous chefs.
One of the toilets is equipped with a monitoring system which automatically analyses urine for the daily control of health.

THE TRON PROJECT

Tron (The Realtime Operating System Nucleus) was the symbol of one of the most prestigious projects of Tokyo University over the last ten years. Its aim was to provide precise indications as to tomorrow's computerized society. The original idea was to manage the evolution of automation in the home to avoid problems similar to those seen with the development of the automobile in a society mostly unprepared for it. The project's starting point was to try to determine what kind of society and environment would foster a better quality of life.

Coordinated by Professor Sakamura, the project carried out a rigorous analysis to determine what kind of interface between various devices and their systems would be most conducive to intelligent objects coordinating their operations instead of working independently.

Sixteen private Japanese companies, including Fujitsu, Hitachi, Matsushita, Mitsubishi and NEC, collaborated to create a computerized home as a kind of experimental laboratory where the project's concepts could be tested and further developed. The pilot house was inaugurated at the end of 1989 in Tokyo's Roppongi district and was used for a series of experiments directly involving people. Data thus gathered mostly concerned conflict between the various control centers and emergency procedures. Fires were simulated, as well as blackouts and other dangerous situations.

Ken Sakamura, Computer Architect, Professor in the Information Sciences Department of Tokyo University, and creator of the Tron project (The Realtime Operating System Nucleus)
"After a long period of overcast skies, a warm spring sun finally pierced through the clouds early this morning and its rays fell on a square house near the center of Tokyo. The atmospheric sensors on the roof register a whole series of data: temperature rise, wind speed, air humidity, the amount of heat reflected off neighboring buildings, even the presence of dust. This data goes through the intricate network of cables that runs throughout the house. The air is certainly warm, even if there is still a slight breeze. In the

abundant sunlight, the house quickly warms up. The southern façade appears to be a large greenhouse because of the glass panels, running from the basement to the roof, which are now silently opening to let in the refreshing outside breeze. The computer has calculated the angle at which the windows should be opened by taking into account all the data provided by the cable network.

Meanwhile the people living in the house are peacefully sleeping.

The central control system has just received confirmation from the atmospheric sensors that the sky will remain clear all day. According to the computer's programming, this means the family will go out for a picnic. So the program set to let them sleep late is canceled and the rolling-shutters go up to let the sunlight into the bedrooms. At the same time, the stereo selects dynamic morning music and the volume is adjusted to create an appropriate atmosphere. The security system goes on stand-by, waiting for the family to get up.

Another day in the intelligent house begins.

Even if it still is not clear exactly why our homes should be automated, home automation appears to have become a reality, so much so that the term 'home automation' was created. A list of domestic functions for which automation would represent great advantages could certainly be made. It is equally true that an automated task could often be performed just as easily by hand, for example, consulting the computer to determine where to put away a pack of crisps. So it is not surprising that automation is often seen as a curious novelty or a kind of game. This does not mean that home automation is only a passing fad with little influence on future home construction. Such an opinion is hardly tenable if one considers the pace at which developments in electronic technology follow one another and their widespread influence on modern society. It is a fact that microchips are already used in homes, even in unexpected places. For instance, they control air-conditioners and many domestic appliances, and one day they will be present in walls, ceilings, and perhaps even in the toilet. If current trends continue, at the beginning of the next millennium, in a few years time, thousands more microchips will be in use and home design will necessarily be affected by the diffusion of the computer.

All this remains relatively insignificant until a way is found to exploit new electronic technology at a global level. Until we have a clear conception of life in the future, what goes under the term home automation will elicit little interest. The risk is we never go beyond aggregations of electronically con-

Tron House.

trolled devices. In the absence of precise ideas to guide us, computers are used only because they are now affordable and because their connection to appliances is technically feasible. As long as these paradigms do not change, it is unlikely the market for home automation will have much success. Putting forward a combined group of electronic functions is not a mistake. However, one also cringes at the thought of an automated rolling shutter controlled by a light sensor which importunately closes and opens, for example while enjoying a magnificent sunset. Houses would become less pleasant to live in. To build something merely because it can be done is unacceptable, as is using it only because it is available. To justify the presence of electronics in a home by appealing to our ability to install them overlooks what a home represents. Automation designed this way should go no further than the drawing board. Since a home is above all a living-space, the first question should be how do people want to live. This is the necessary point from which to start imagining domestic automation systems. If the criterion remains the feasibility of an installation, the future looms grim indeed. Automation and the use of computers only make sense in regard to real needs. Besides, electronic technology has reached a state where just about anyone's fantasies can be realized. The only obstacle remains the cost. It should be noted, however, that small computers which cost a few hundred thousand yen are more powerful than large computers which, 10 years ago, cost a hundred million yen. The problem of cost often solves itself as soon as demand increases. More foresight is required. We must ask ourselves how we would like to live. This is the perspective which can lead to the use of automation in the home, and this was the guiding concept for the design and construction of the Tron house. The main reason automation should be brought to the home is to make life easier. However, the aim is not to build houses for lazy people: there is no need of electronics for that. First of all we must consider what computers are. The image which occurs to me is that of little assistants waiting for an occasion to help us, a little like servants at a king's beck and call, ready to perform tasks he might easily have carried out himself. Thus, and this is essential, computers can help people who can no longer manage alone.

Let us suppose that everything in a home is automated. Without this basic idea of offering assistance to people, someone who had enjoyed taking care of their plants over the years might suddenly find themselves ousted by a computerized watering system. The negative aspect of this example is due to the invasion by the computer of the area of human sentiment. But we must not forget that in addition to people who love watering their plants, some people, elderly or disabled, cannot do this.

In other words, a good domestic automation system must adapt itself to people, allowing people to water their plants alone or leave this task to the computer or take care of them themselves but with the computer's assistance, for instance, in using the correct amount of water. There must be no constraints, but rather more and more choices. Moreover, an activity such as watering plants requires not only physical ability but time as well. Many people who would like to surround their home with plants do not have the time to take care of them. There are also those who, due to illness or an accident, are temporarily disabled and unable to carry out the simplest tasks, including the pleasant ones.

This is the reason why home automation, seen as help in the home, is not considered merely as assistance for people with particular needs such as the elderly or disabled, even though fitting out homes for them is certainly a way to develop flexible systems for different domestic scenarios. Keeping people with special needs in mind is a good way to work towards home automation since it both helps these people and enables new solutions to be found. Until home automation is developed from this perspective, we will remain in the blind alley of self-gratification.

What is important is that anybody, literally anybody, must be able to use an installed system. Computer circuits are now found in all kinds of objects and, more and more, we will have to involve ourselves with electronics. But this must happen because it is necessary for everyday life just as light-switches are necessary today. Children, the elderly, people who do not usually operate machines and appliances, the disabled, everyone will have to be able to use such apparatuses otherwise their life will become even more difficult than it is today. The present approach towards home automation must be revised. There are systems which automatically close windows and activate the air-conditioner as soon as the external temperature rises. In my opinion,

Checking panel.

they should instead open the windows wider to let in a possible refreshing breeze. This is the kind of automation that should be aimed for. We are now arriving at the crucial point of the issue which is similar to the situation in which, for example, a person does not want to cook because they find it boring, so they turn to pre-cooked or instant products. It would make more sense to use an automated system which is able to cook with fresh products. Integrated circuits today are able to make possible precisely this kind of home automation based on a wider choice of options. In a private home, electronic circuits are found in the air-conditioner, ventilation system, microwave oven, refrigerator, washing machine and generally speaking all electric appliances. Without our having paid much attention to the amount of electronics they contain, these objects have become quite common. Such devices could be called intelligent objects and their number is going to significantly increase in the next few years. We will come to live in 'computerized cities', urban conglomerations brimming with electronics. However, the presence of computers will not be the most important feature. The main characteristic will be the connection of all these machines by means of computerized networks. Connected electric circuits will interact and coordinate their functions. If one of these needs help to complete a certain task, it will be able to contact others for assistance and information. If any conflicts arise between equally viable but different solutions, they will probably reach compromises. The combination of all available data will lead to new services which the single installations, operating independently, could not have offered. Inside private homes, data gathered by the security system will be transmitted to the lighting system so that the lights go on and off as a person enters and leaves a room. Similarly, data concerning the strength of sunlight will be used not only by the lighting system but by the air-conditioning which will also take into consideration the amount of heat radiating from electric appliances. You could also imagine that the air-conditioning, warned in advance of an imminent pianissimo passage in the classical piece being played on the stereo, would work out a compromise between maintaining the requisite temperature and operating so as not to drown the music.

A system such as the one just described represents a distributed system where each electronic circuit, from the simplest to the most complex, is able to interact with the others so as to foster an environment most adapted to human needs. A distributed system is more trustworthy since a malfunction in one of the parts does not affect the functioning of the entire system. It also better lends itself to modifications such as the addition or removal of equipment or the transformation of the distributive plan. Furthermore, cabling is simpler than with a centralized installation, and services are cheaper. Let us call such a system a 'highly functionally distributed system' (HFDS). In an HFDS, the relation between the different parts, their physical arrangement in space, is extremely important. A single HFDS can constitute one component of a larger system. For example, a group of intelligent apparatuses and objects could constitute a system in a room, a group of room systems could form a domestic system, several domestic systems could represent a building system, a group of building systems could form a local system, then a district system, city system, and so forth. The final limit is a global system.

Considering the specific impact which advanced systems will have on homes, it is important to bear in mind the present trend from centralization to distribution. A truly functional 'distributed system' would represent the final consequence of this trend. When our cities are equipped with computerized networks and high-speed communication lines, city services, both qualitatively and quantitatively, will be accessible everywhere. Above all, it will be possible to react to each situation in real time. What happens when this concept of functional distribution is transferred to the home? A home contains much space to put things away such as chests and cupboards which simple automation does not reduce in size. The interesting point is that most objects put away are used very infrequently, Christmas tree lights for instance. It does not make any sense to accumulate these objects especially considering real estate costs in cities. One solution would be to store such objects in underground computer controlled warehouses. If we needed something, all we would have to do is ask the computer that would know where to go to bring it to us. The computer could also draw our attention to objects we no longer use and send them, upon our prompt, to a retailer. Furthermore, production could also be distributed. In terms of agriculture, it might be desirable to grow crops near their place of consumption so as to be supplied with fresh products. Many vegetables are already cultivated on the outskirts of large cities, but what I have in mind is the possibility of growing them inside our own homes. It is not a question of transforming building courtyards into fields. Present electronically assisted hydroponic cultivation enables plants to grow with neither soil nor chemical pesticides. Even if it may seem rather unnatural to grow and harvest vegetables all year round, it is in fact the only way to savor their true taste and, for the city-dweller, to directly obtain the tomatoes, asparagus, lettuce, and herbs he or she usually consumes. Electronics can assist plant growth so that people do not have to lift much more than a finger. Energy and fertilizers derived from organic sources can be used while the heat and refuse can be recycled.

Once again, the computer plays a fundamental role. Distribution was essential in the past when scattered communities produced that which they needed to survive. Then centralization appeared with large scale production, since the previous model proved ineffective when faced with new requirements. It is not a question of turning back. However, distribution can become interesting again thanks to the computer whose constant control can make small scale production valid again. Thanks to the interconnection between all installations, it is possible to make very specific use of the service combination possibilities. Finally, it is worth observing that, before it is built, the computerized city will require a different organization than that existing in present cities. We have seen that electronic systems can represent the antithesis of centralized authority and rigidly structured administration, even though computers can be used for such aims. In an age of individuality and diversity, demand for functional distribution and specialization would grow. The role of electronics would then be to enable the distribution of services and functions in time and space, and promote diversification, by maintaining interconnections throughout this expansive complex."

"Domotic module" control.

Among the towers of the La Défense business district in Paris, the Bull Tower presents interesting characteristics in terms of technical installation management.

The building is used as the French headquarters for all the business services of Bull. The intention of the sponsor was to provide geographic coherence to a group which unites various firms.

The building was designed by the architects Andrault and Parat, noted for their work at Bercy and their reconversion of the nearby CNIT dome at La Défense. The tower has an elliptic shape and an overall floor space of 65,000 square meters distributed over 33 floors and accommodates 2,100 permanent employees. The mezzanine floor and the next two serve as operations center and product demonstration rooms for clients. The remainder is used as office space. A spectacular suspended shuttle directly connects the tower operations center and the Bull showroom in the Infomart space beneath the concrete CNIT dome.

Vertical transportation is provided by 16 lifts, 8 for the 19 first floors, and 8 which go straight to the remaining floors.

The building's most notable feature is the fact that all technical networks are pre-cabled and run along a single conduit beneath the floors on each story. There are no pipes in the walls and these are made from modular elements; easy to move according to need.

Three cables come out of the units on the floor in each room. The first is for electric power supply, the second for voice and data that goes to the phone and computer, and finally the last is connected to a little control box, barely larger than a cigarette pack. Employees call this device a "domotic module" and it replaces usual switches by enabling the occupants of each room to regulate the internal conditions according to their requirements.

For example, the air-conditioning can be turned on and off in each individual office by means of three buttons identified by the red and blue triangle which surrounds them. A cursor on the side enables the prefixed temperature to be modified by plus or minus three degrees, levels by no means negligible in an office, especially compared to what is usually made possible by more traditional systems. The module also enables control of the fluorescent lights in each office. For this, two buttons are used which are easily recognizable because of the yellow circle surrounding them.

Again using the small remote control box, it is also possible to regulate the shades enclosed in the double-glazing of a maximum of four windows per office. Four rows of three buttons surrounded in green are used for this. Top buttons raise the shades, bottom ones lower them, and middle ones stop them.

The Bull Tower is equipped with an automatic system for the conveyance of documents and post. Documents are enclosed in containers and coded so that the automated system can deliver them via pneumatic tube to the floor of the recipient whose domotic module has a special signal that indicates correspondence has arrived for him or her.

Perhaps the most interesting feature is that the remote control does nothing but emit signals to a special network which is entirely distinct from the central integrated network controlling all the systems and installations in each zone.

The special network conveys the signals to special technical rooms on each floor where all such activation takes place. It is in these service rooms that, for instance, air temperature and ventilation are regulated according to the wishes of each employee. It is also here that all switches are situated, from light switches to the ones for the shades.

From the point of view of system management, this means that all mainte-

Bull Tower in Paris: technical room on each floor.

nance and repair takes place only in these rooms. If one of the air-conditioning pipes is leaking, the technician never need intervene in the office but can work in the special service room without fear of disturbing anyone. The only operation actually taking place in the offices themselves is light replacement. A centralized and automated system takes care of the building as a whole, covering all security and management aspects. For the offices, for instance, it is responsible for the treatment of air during work hours, turns off the lights during the lunch break and at night, and lowers the night shutters.

Such operations do not prevent employees from continuing their work since it is always possible to turn the light back on with the module, just as the shades rise in the morning when employees arrive at the office.

Eventually, a compromise is reached between the need to reduce general energy consumption and an attempt to better accommodate each person's preferences. This goes hand in hand with the new trend, seen especially in Europe, for corporate sector buildings to develop sensitivity to individual needs.

WTC in Stockholm

The WTC building in Stockholm covers an overall surface of 85,000 square meters and is about 30 meters high. It houses the World Trade Center offices, shops, restaurants, parking lots, and the City Terminal, a multi-story complex for national and international train and coach lines, including the airport shuttle. Every day, more than 25,000 people go through the building.

The complex actually consists of 8 different buildings with steel skeleton structures and ferroconcrete ceilings. The complex is brought together by a glass dome measuring 10,000 square meters, one of the largest in Northern Europe.

The dome forms an internal gallery 300 meters long and 25 meters high which is the central space where all activity converges. This roof forms an integral part of the heating and cooling system which, though not particularly complex, was conceived not only for energy self-sufficiency during most of the year, but also to supply Stockholm's district heating system with any excess energy. In fact, the complex only needs to purchase around 15% of the energy it consumes each year. This is mainly due to the winter months when temperatures often drop far below zero, though the center is able to accumulate and produce energy even with external temperatures of minus

10 degrees Celsius. Energy supply to Stockholm is more consistent however and concerns about 2 fifths of the overall energy budget.

Heat is provided mainly from the glass dome, but also from lighting in offices and public spaces, as well as from the presence of people. The building was conceived to be able to trap heat and maintain it inside as long as possible. During construction, particular attention was paid to the choice of materials to make certain they provided maximum thermal insulation both individually and in combination with those surrounding them. Insulation reaches such levels that offices must be cooled during working hours even in the winter. In terms of energy management, air-conditioning installations with both natural and artificial circulation are combined with a fan-coil installation. Both systems are connected through heat exchangers which heat or cool the relevant fluids according to need. The basic design provides for hot air to be extracted through openings beneath office windows and sent to the central cooling system where heat is distributed through the fan-coil system or, with excess heat, sent to urban district heating. Four large ventilators placed near the dome ensure air is redistributed in the public spaces and conveyed to those exchangers which recover any energy surpluses.

All heating and cooling operations in offices and public places are controlled by a central computer system whose functions are integrated with the communication and security systems of the complex.

5. Transparent Outer-shells

Glass plays an important role today in the outer surfaces of buildings. They now carry out an ever growing series of tasks which would have seemed inconceivable 20 or 30 years ago. The transparency and capability of fenestration to present views without optical distortion are now well known characteristics taken for granted. Other features concern levels of luminosity, air exchange inside the building, thermal and acoustic insulation, and safety, both in the sense of safety for the user and security against burglary. Requirements have become more and more specific. To meet them the "fenestration-glass curtain wall" system has developed a series of sophisticated devices. These range from shape changes in frames, to double or triple glazed panes with empty spaces in between, to the use of plastic films which both increase resistance and reduce heat loss and the strength of sun-light.

Mostly, these are "passive" and in many cases non-modifiable solutions which have brought about improvements in the area of fenestration that, though important, have not however been able to fully respond to ever diversifying requirements.

A window does not yet exist which, though remaining transparent and enabling the occupants to see outside, is able to automatically modify some of its qualities in response to weather conditions or security needs.

It is, however, possible to identify in the last ten years a few interesting and even renowned prototypes of this new product which we could call "intelligent windows" even if the term "intelligent" has been a little abused recently. The aim of this far from exhaustive review is only to demonstrate that intelligent windows are not too far away in the future.

One of the first pioneers, as early as the mid-1970s, was the American Steve Baer, director of Zomework Corporation. He is most known for his house with a geodesic asymmetrical dome and walls made of superimposed metal containers that form a gigantic solar panel.

His company built a peculiar greenhouse for the elementary school of Monte Vista in New Mexico whose south-facing, multi-

Toronto Hydro Place.

layered glass roof formed an integral part of the solar heating system.

This special roof consists of two sheets of transparent Plexiglas through which sunlight enters to warm the inside, while heat is accumulated in containers filled with water placed against the wall at the back.

At night, or when the weather is overcast, minute particles of insulating material are automatically pumped between the two glass layers and completely fill the space between them thus forming an effective insulating surface.

A team of Canadian experts, under the name Encon, who [76] worked at the end of the 1970s sought to integrate glass surfaces into the heating and air-conditioning system at a sophisticated technological level.

The most interesting solutions stem from extremely simple observations and concern the windows themselves. They noted especially that heat is never homogeneously distributed inside a building.

There are colder areas near the walls and windows while heat remains trapped in the center. This is why central areas of large business buildings must often be cooled even in the winter by means of air-conditioning. So a system was developed which, instead of using traditional heating methods, channeled excess heat radiated from devices, lighting, and people and circulated it around peripheral areas, storing any superfluous heat in special reservoirs.

The two buildings built according to these results are extremely sophisticated, even for today's standards, especially as regards their insulation solutions. The entire perimeter of the building is "padded" by means of double-glazed fenestration with an outer surface which reflects more than 85% of sunlight.

This glass wall is conceived as a single installation along with the false ceilings above the offices. Between the two glass layers, conduits are incorporated for the distribution of hot air which is re-introduced into the building through grids whose placement causes a draught along the glass.

Basically, it is the same system as that used to defrost a car windshield. Air circulation along the surface assists in maintaining a

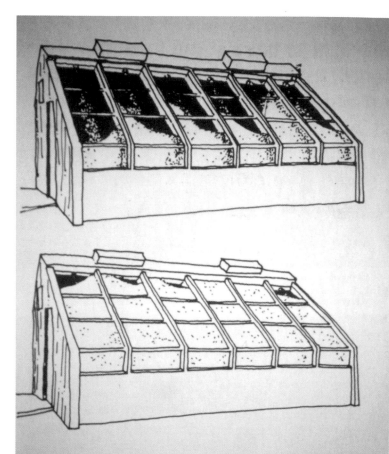

AT SUN SET BEADS OF INSULATION ARE
FORCED INTO THE PANELS TO CONTAIN THE
SOLAR ENERGY GENERATED DURING THE
DAY.

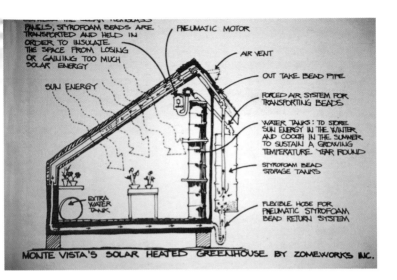

Above and facing page: Steve Baer, Monte Vista school, New Mexico.

certain temperature. Thanks to this device it is possible to recover 4 to 10% of those spaces whose proximity to the windows made them previously unusable since they were always either too hot or too cold. In the false ceilings, along with all other system networks, there are openings for hot air intake as well as reservoirs for excess hot air and heat exchangers which heat or cool the air before it is re-introduced into the work-place as described.

A central computer continuously controls the climate conditions within the building, and analyses energy intake and output as well as storage levels.

The system was tested in 1980 in two important Canadian business buildings: the Hydro Place in Toronto and Gulf Canada Square in Calgary. Energy savings remained constant at around 40% in the first building and 35% in the second. [12]

The next important event in the history of the intelligent window was the no longer functional façade of the Centro Internazionale

Encon System Diagram.

Milan Fair, CISI façade.

degli Scambi in Milan, completed in 1981 as part of the *Milan fair* 77-78
energy saving plan.

It was the first experiment using an outer-shell with variable
transparency. The objective was to avoid non-modifiable solutions, such as the essentially passive Trombe wall, and find a
solution which exploited the free circulation of sunlight to meet
energy requirements.

Accordingly, the façade was divided between areas with solar
panels to heat water for sanitary use and variable transparency
areas able to regulate the entry of sunlight. Solar energy was
controlled using the properties of transparent catadioptric elements which admit light with a slight deviation or reflect it at 180
degrees depending on the presence or absence of water.

One hundred of the façade's 250 modular panels were variable
transparency panels composed of triple-layered glass such as

Milan Fair, CISI façade. Above: variable transparency windows. Below: screen with reflecting surfaces.

thermophan with a Plexiglas screen inserted inside. The latter was stamped to created minute reflection prisms (catadioptric elements similar to those commonly used on bicycle tail lights).

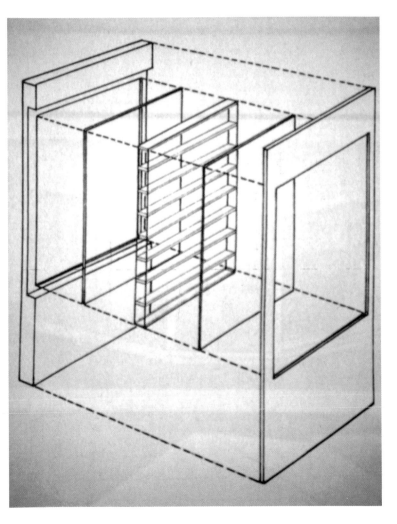

Illinois Institute of Technology, diagram of window installation.

Institut du Monde Arabe, Paris.

Institut du Monde Arabe, Paris, photoelectric cell in the roof, controlling the entire system.

When these prisms are inclined, more than 90% of solar light is reflected. When water is introduced between the catadioptric elements, the panel becomes semi-transparent and admits 75% of sunlight. The view of the outside, however, was not clear because of the presence of water and methacrylate.

Since the introduction of water could be regulated, a "curtain effect" could be obtained which only let in as much light as desired. A depressurization device with a valve controlled water circulation and compensated for hydrostatic pressure on the glass to avoid breaking, while an insulating layer of air between the catadioptric elements and the external glass prevented thermal dispersion in the windows full of water during cold periods. Since the panels never actually functioned all together, it was not

Institut du Monde Arabe, Paris, interiors.

possible to collect and process operation data so as to implement automated management of the façade.

Developed by design students of the Illinois Institute of [79] Technology in Chicago as part of their degree program, the "House of the Future," was better geared towards the market than the Milan building. It was not actually a project but rather an aggregation of separately evolved concepts brought together in the idea of a house of the future. A technique called "structured planning" was used to find solutions for particular problems, e.g. home security, with qualitative information. In the specific case of anti-theft, the exterior windows represented an interesting example.

Apart from being transparent walls they formed an integral part of the security system. In *the House of the Future* there were many options in terms of fenestration and the future owner had a large choice to select from.

The windows were made of plastic material to reduce their weight. Within them, however, two panes were separated by an empty space to insure good insulation. A shading system was also incorporated into the glass. It consisted of blades like a Venetian blind which slid both vertically and horizontally and which could be automatically activated by the central system which received temperature information from internal and external sensors. Each side of the blades had a different profile so as to absorb or reflect sunlight and a transparent film covered the glass itself to increase security and resistance to abrasion. The use of plastic materials and laminated glass made the windows very rigid and difficult to break into.

The windows were also conceived as an integral part of the anti-intrusion security system. A pneumatic sensor was placed between the double-glazing which was able to record each increase in pressure within the partially empty inner chamber. An increase in pressure above a certain threshold indicated an attempt to break through the glass wall and the central security system was activated. In order to ease maintenance, the windows were mounted on hinges to rotate 180 degrees to give effortless access to both sides from inside.

There were three possible fenestration sizes, full-opening, three-quarters, and half. These were available both in vertical and horizontal versions.

One of the most important stages in the evolution of glass
80-83 facades was the Institut du Monde Arabe built at the end of the 1980s in Paris. It was completed in 1987 in one of the few open areas in the city center and constitutes one of the most interesting combinations of architecture and technology, especially with its glass curtain wall able to meet the building's environmental requirements.

This particular façade is on the south wing, an enormous solar screen measuring 30 by 80 meters which regulates the admission of light and creates complex geometric patterns inside. It is composed of 240 square panels measuring 2.40 square meters

Lyon Panorama, window frame with sensors.

and suspended in a metal framework. Each panel is made of three layers of transparent glass (double glazed outside and simple glass inside) and contains a screen plate perforated with 73 holes. The largest is in the center of each panel and consists of a mobile aluminum diaphragm surrounded by 56 fixed or partially mobile smaller diaphragms. There are more than 27,000 such diaphragms covering the façade. Because of them it is possible to control the quantity of light admitted inside thus creating an endless play of lighted and shaded geometric shapes.

The mobile diaphragms are essential for this. They function very much like those found in cameras. They are opened and closed by means of mobile stainless steel rods set in motion by two electric motors.

Movements of the mechanisms on the façade's panels are automatically controlled by a photoelectric cell placed on the building's roof, but they can also be manually remote-controlled. Depending on the higher or lower presence of light, a signal is emitted to open or close the diaphragms.

The completed building with its simple control system is very different from the initial design which planned for a much more sophisticated system of light management with a large number of electronic sensors on each panel to automatically and continuously determine the opening of the diaphragms with intermediate positions depending on the intensity of external light.

One of the vertical sides of each square panel is hinged so that the façade can open like a large window. For purposes of maintenance and cleaning, the inner glass layer is also hinged.

In the field of variable transparency, mention must also be made of so-called "electronic windows" capable of modifying their own light-refraction properties, from perfect transparency to total opaqueness, as well as changing color, according to external weather conditions.

Two procedures make such results possible. First, the glass or polycarbonate sheet contains a very thin layer of encapsulated liquid crystals. These crystals, called Nematic Curvilinear Aligned Phase (NCAP), position themselves equidistantly on the surface of microscopic transparent polymer spheres. The latter are assembled in a colloidal laminated matrix between two fine transparent plastic films whose surface is treated to make it electrically conductive.

In the absence of electric tension the casually scattered spheres with the liquid crystals prevent the passage of light through the film so that the glass sheet has the appearance of smoked glass. When an alternate current passes through the two surfaces of the film, an electric field is created whose strength surpasses the weak force of the polymer spheres. The liquid crystals are thus aligned along pre-determined lines and the glass becomes transparent since the light rays are no longer intercepted by the crystals and can now traverse the film.

The second procedure is based on the action of an electric discharge, either positive or negative depending on what the aim is, which stimulates two compounds contained between glass sheets: lithium trioxide and tungsten trioxide. Depending on the valence of the discharge, these compounds mix in a certain way and modify the color and transparency of the windows. External sensors determine the level of light and emit, by means of an electronic regulator, the impulse responsible for the electric discharge.

The system provides standard settings of transparency and opacity as well as the possibility of carrying out direct adjustments to determine the final positioning of the windows according to the specific conditions.

Producers of window accessories also supply windows and systems used to manage a building's fenestration; i.e. the opening, closing and shading of the windows, from a PC.

Such systems were originally conceived to help the disabled to open and close their windows, but their field of application subsequently grew to cover all fenestration functions in new and already existing buildings. Special sensors placed outside the building, and able to determine the degree of light and humidity in the atmosphere, send signals to the central computer which analyses the data using a standard program and issues commands to activating mechanisms to adjust, open, or close the windows.

Similarly, it is possible to regulate shading by positioning the blades contained inside the double-glazed panes according to the time of day, temperature and angle of sunlight. A timer even makes it possible to preset the hour when a certain task should be performed, for instance, to air out a room taking into account the hour at which noise levels outside are highest. Since the position of the windows is continuously controlled, the system can also be used to identify break-in attempts.

It goes without saying that, as a necessary precautionary measure, each window can always be opened directly by infrared remote control as well as by hand.

The operation of such systems has been tested for ten years in a
85 few experimental buildings such as the showcase house *Lyon Panorama* completed at the end of the 1980s by Electricité de France (EDF), the corporation which manages French electrical energy.

Among other things, this building, no longer used today, was provided with an interesting system of intelligent glass walls. Two sensors were placed inside the white PVC windows near the handle. These were connected to the centralized computer through the security system network. The first sensor's task was to send a signal when the closed window underwent strong pressure from the outside.

Whenever a window remained open beyond a certain length of time without any signals from the perimeter protection system, for example when the room was being aired out, the second sensor sent a signal to the system controlling energy usage which, if the heating was on, turned off the room's heat until the window was closed again, at which time the temperature was raised again to the required level.

A third sensor was positioned outside the window on the slide of the rolling shutter and, reacting to undue pressure, was used to indicate possible intrusion attempts. As regards centralized control and management, there were in fact various subsystems, each with its own tasks, and all interconnected.

The idea of "intelligent windows" goes hand in hand with the general tendency to include more and more sophisticated management and control systems in buildings and with the massive introduction of sensors and integrated electronics in everyday objects.

The development of specific technology necessary for such products as well as their being taken up by the market will depend, as always, on whether or not the products are easy to use and install and the availability of the necessary resources. In any case, this new product will follow its own course and has an extremely interesting role to play in the present market of outer fenestration.

6. Considerations on High-technology in Everyday Life

The world we live in is taking shape as a heterogeneous complex of high-tech devices. Ever more powerful and sophisticated machines are now a permanent presence not only in space but also in our urban and domestic environment. The common characteristic of the objects surrounding us from the telecommunication satellite to the latest model of a multi-programmable microwave oven is in fact their great complexity manageable only by means of elaborate electronic control devices. Thanks to this disseminated intelligence, a piece of equipment operated today is no longer inanimate as in the past. It sends signals, communicates in its own idiosyncratic manner and enables people to interact with it in a dynamic way. Most people do not know how these objects function and are content to know enough about their operation to be able to perform the few functions which interest them. How many people actually know what happens inside their phone when they lift the receiver and start dialing? A survey on American domestic habits carried out at Harvard University has revealed that four fifths of the interviewees thought that signals were emitted in order to reach a certain phone number but could not understand how their friend's voice, with its characteristic inflections and timbre, could pass through the cable.

Greater dependence has gone hand in hand with the increase in complexity. Thus, in order to operate any piece of equipment, we are totally dependent on extended networks. Setting aside the telephone and Internet as all too obvious, let us consider the washing machine, present in more than 97% of Italian homes, as an example taken among domestic appliances. Not only must it be plugged into the electric socket, but it must also be connected to the water supply. Furthermore, each wash requires detergents and softeners, obtainable only through the network of major consumer goods. If one of these structures failed us, we would be left with our dirty laundry.

Since its internal workings are little understood, the day-to-day experience of using high technology differs radically from what designers and producers might have expected or intended. In this regard, security devices and installations offer a good starting point for further reflection, precisely because their characteristics will enable us to shed more light on the relation between users and technological products. A typical case concerns the field of security. The Jones', worried about recent burglaries in their friends' homes, finally resolved to have an alarm system installed in their home. At first everything worked fine. Then the alarm started going off. The first time was during a storm, then because of a leaf, a draft, and then finally "for no reason at all", or at least for no reason which the technician could identify. According to researchers of the French CNRS team investigating new technologies and transformations in life-style, if a problem occurs more than couple of times, the individuals' psychology starts coming into play, as well as the manner in which they relate to others and their environment. It is a striking fact that people commonly resort to believing that their security system, provided with its own electronic intelligence, goes off to "play a trick" on its owner. Our everyday world now abounds with devices that look for the earliest opportunity to malfunction, from the car which will not start because its battery is dead even though you never forgot to turn the lights off, to the videocassette player that records TV adverts precisely on the cassette with your favorite film. The case of the Observer space probe caused a sensation in 1993 when it malfunctioned, never revealing the secrets of Mars, and, as American newspapers reported, caused three quarters of the Pasadena technicians involved in the project to be sent to psychiatric hospitals, convinced the satellite had conspired against them.

There is however a fundamental difference between objects that do not function for an unidentified reason and systems like alarms which go off without any apparent cause. The latter are guilty of operating excessively, they are too eager to perform the task for which they were conceived. However, an alarm does not go off because if feels like it but because its internal components and programs were conceived precisely to that end. In fact it is installed for that very purpose and users expect it to indicate the presence of a burglar, or anyone taken for one, in the most conspicuous way. Security system operators say half jokingly that their customers would like their alarm system to be able to physically arrest possible thieves even before the police arrive. In other words, people will readily accept that their system goes too far in certain cases but not in all cases. Not for a draught which at worst may damage your health, and certainly not for a leaf whose fall most likely will not be dangerous. Man-machine relations are rather peculiar in this instance. The security system is expected not only be able to exactly evaluate a situation in each particular case, but also to interpret the feelings of individual people. According to the researchers at the Center of the Sociology of Innovation at the Ecole des Mines in Paris, French installers often face two types of psychologically motivated requests. The first is for the alarm to go off only if the intruders are "clearly" burglars. The other regards the possibility of keeping the system disconnected to avoid all risks of malfunction as in the case of the Jones', allowing however for the system's "self-activating" in case of a clear break-in. There's a sense that these people are frightened not so much of burglars as of their own alarm. They seek to protect themselves no longer with but from their security system. The more sophisticated the technology, the more users become mistrustful. Modern alarms are not limited to just sounding an alarm. In fact they often no longer emit sound signals at all. They are now connected to many recorded numbers by means of the phone. Surveillance services, police, the office, relatives, all are simultaneously contacted in a case of presumed danger. This triggers a process involving many people who trouble themselves perhaps for nothing. This becomes difficult to manage for the owner of the alarm system, not so much because of the excuses he must present to all those concerned, but on account of his uncertainty that they will react in the future, precisely perhaps when their intervention is actually necessary. From this point of view, there appears to be no way out apart from falling back on the usual explanation: the machine is "spiteful" and cannot stand the owner. That this is the most frequent interpretation says a lot about the degree of alienation pervading our life.

If however the user and the device are no longer opposed to each other but seen in connection with the market of security devices, worth more than a trillion lire in Italy for the systems alone, a fundamental observation must be made. There are various players in this market. First of all, the inventor(s) and producer of the system, commonly technicians able to understand the construction and operation of machines. Then come the business people, those responsible for marketing and distribution, and the salesmen and installers, particularly interested in how much a product can be adapted and tailored to suit the needs of the customer. Each of these is concerned with his own issues and objectives, often quite different from those of the others. No machine was ever transferred from the technical sphere to business and sold directly. There are continuous modifications with functions and components removed or added so as to arrive at a final product, which will never reach absolute perfection since no one would presently be able to afford it.

The user stands at the other end of the market. He is very much attracted to high technology presented as it is today as the most practical aspect of scientific theory. No matter what values are invested by the owner in his recently purchased device, he cannot dissociate the product from what it does. Just as a video recorder is designed to record images on a tape, a security system must emit an alarm. The problem is that we are still far from absolute perfection, which, among other problems, is difficult to combine with competitive prices. It therefore happens that malfunctions occur. In such cases the consumer may realize that he is not the privileged owner of a product conceived expressly for him as TV adverts may have led him to believe, but more simply the final part of a long process in which he is only one of the players, even if he must pay to participate.

The Information Technology Revolution in Architecture is a new series reflecting on the effects the virtual dimension is having on architects and architecture in general. Each volume will examine a single topic, highlighting the essential aspects and exploring their relevance for the architects of today.

Other titles in this series:

Information Architecture
Basis and future of CAAD
Gerhard Schmitt
ISBN 3-7643-6092-5

HyperArchitecture
Spaces in the Electronic Age
Luigi Prestinenza Puglisi
ISBN 3-7643-6093-3

Digital Eisenman
An Office of the Electronic Era
Luca Galofaro
ISBN 3-7643-6094-1

Digital Stories
The Poetics of Communication
Maia Engeli
ISBN 3-7643-6175-1

Virtual Terragni
CAAD in Historical and Critical Research
Mirko Galli / Claudia Mühlhoff
ISBN 3-7643-6174-3

Natural Born CAADesigners
Young American Architects
Christian Pongratz / Maria Rita Perbellini
ISBN 3-7643-6246-4

New Wombs
Electronic Bodies and Architectural Disorders
Maria Luisa Palumbo
ISBN 3-7643-6294-4

Digital Design
New Frontiers for the Objects
Paolo Martegani / Riccardo Montenegro
ISBN 3-7643-6296-0

The Architecture of Intelligence
Derrick de Kerckhove
ISBN 3-7643-6451-3

Aesthetics of Total Serialism
Contemporary Research from Music to Architecture
Markus Bandur
ISBN 3-7643-6449-1

For our free catalog please contact:

Birkhäuser – Publishers for Architecture
P. O. Box 133, CH-4010 Basel, Switzerland
Tel. ++41-(0)61-205 07 07; Fax ++41-(0)61-205 07 92
e-mail: sales@birkhauser.ch
http://www.birkhauser.ch